Editor
Gisela Lee

Managing Editor
Karen Goldfluss, M.S. Ed.

Editor-in-Chief
Sharon Coan, M.S. Ed.

Art Coordinator
Kevin Barnes

Art Director
CJae Froshay

Imaging
Alfred Lau
James Edward Grace

Product Manager
Phil Garcia

Publisher
Mary D. Smith, M.S. Ed.

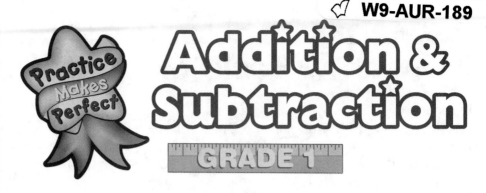

Addition & Subtraction
GRADE 1

Authors

Teacher Created Resources Staff

Teacher Created Resources, Inc.
6421 Industry Way
Westminster, CA 92683
www.teachercreated.com

ISBN-0-7439-3315-X

©2002 Teacher Created Resources, Inc.
Reprinted, 2006
Made in U.S.A.

Table of Contents

#3315 Practice Makes Perfect: Addition and Subtraction
© Teacher Created Resources, Inc.

Introduction

The old adage "practice makes perfect" can really hold true for your child and his or her education. The more practice and exposure your child has with concepts being taught in school, the more success he or she is likely to find. For many parents, knowing how to help their children can be frustrating because the resources may not be readily available. As a parent it is also difficult to know where to focus your efforts so that the extra practice your child receives at home supports what he or she is learning in school.

This book has been designed to help parents and teachers reinforce basic skills with their children. *Practice Makes Perfect* reviews basic math skills for children in the first grade. The math focus is on addition and subtraction. While it would be impossible to include all concepts taught in the first grade in this book, the following basic objectives are reinforced through practice exercises. These objectives support math standards established on a district, state, or national level. (Refer to the Table of Contents for the specific objectives of each practice page.)

- one-digit addition
- one-digit addition with regrouping
- addition facts to 20
- one-digit subtraction
- number line usage

- subtraction facts to 10
- word problems
- two-digit addition and subtraction with regrouping
- two-digit addition and subtraction without regrouping

There are 36 practice pages organized sequentially, so children can build their knowledge from more basic skills to higher-level math skills. To correct the practice pages in this book, use the answer key provided on pages 47 and 48. Six practice tests follow the practice pages. These provide children with multiple-choice test items to help prepare them for standardized tests administered in schools. As children complete a problem, they fill in the correct letter among the answer choices. An optional "bubble-in" answer sheet has also been provided on page 46. This answer sheet is similar to those found on standardized tests. As your child completes each test, he or she can fill in the correct bubbles on the answer sheet.

How to Make the Most of This Book

Here are some useful ideas for optimizing the practice pages in this book:

- Set aside a specific place in your home to work on the practice pages. Keep it neat and tidy with materials on hand.

- Set up a certain time of day to work on the practice pages. This will establish consistency. An alternative is to look for times in your day or week that are less hectic and conducive to practicing skills.

- Keep all practice sessions with your child positive and constructive. If the mood becomes tense, or you and your child are frustrated, set the book aside and look for another time to practice with your child.

- Help with instructions if necessary. If your child is having difficulty understanding what to do or how to get started, work the first problem through with him or her.

- Review the work your child has done. This serves as reinforcement and provides further practice.

- Allow your child to use whatever writing instruments he or she prefers. For example, colored pencils can add variety and pleasure to drill work.

- Pay attention to the areas in which your child has the most difficulty. Provide extra guidance and exercises in those areas. Allowing children to use drawings and manipulatives, such as coins, tiles, game markers, or flash cards, can help them grasp difficult concepts more easily.

- Look for ways to make real-life application to the skills being reinforced.

Practice 1

Count the items in each box. Write the sums.

1.

$\underline{\hspace{1cm}} + \underline{\hspace{1cm}} = \underline{\hspace{1cm}}$

2.

$\underline{\hspace{1cm}} + \underline{\hspace{1cm}} = \underline{\hspace{1cm}}$

3.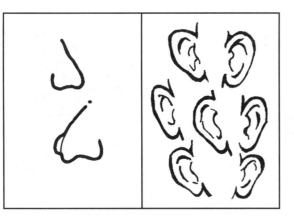

$\underline{\hspace{1cm}} + \underline{\hspace{1cm}} = \underline{\hspace{1cm}}$

4.

$\underline{\hspace{1cm}} + \underline{\hspace{1cm}} = \underline{\hspace{1cm}}$

5.

$\underline{\hspace{1cm}} + \underline{\hspace{1cm}} = \underline{\hspace{1cm}}$

6.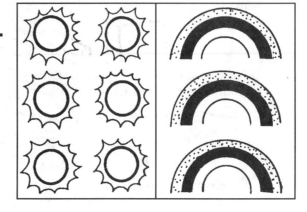

$\underline{\hspace{1cm}} + \underline{\hspace{1cm}} = \underline{\hspace{1cm}}$

Practice 2

Count the items in each box. Write the sums.

1.

$$2 + 2 = \underline{\hspace{1cm}}$$

4.

$$\underline{\hspace{1cm}} + \underline{\hspace{1cm}} = \underline{\hspace{1cm}}$$

2.
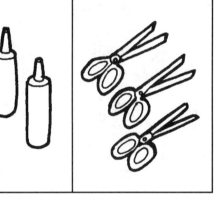

$$\underline{\hspace{1cm}} + \underline{\hspace{1cm}} = \underline{\hspace{1cm}}$$

5.

$$\underline{\hspace{1cm}} + \underline{\hspace{1cm}} = \underline{\hspace{1cm}}$$

3.
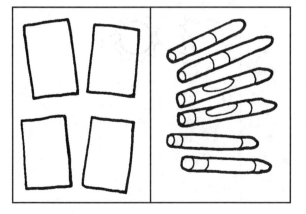

$$\underline{\hspace{1cm}} + \underline{\hspace{1cm}} = \underline{\hspace{1cm}}$$

6.
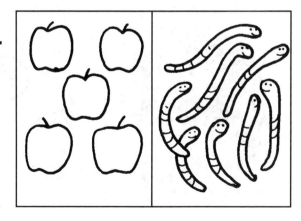

$$\underline{\hspace{1cm}} + \underline{\hspace{1cm}} = \underline{\hspace{1cm}}$$

Practice 3

Count and then write a number sentence.

1.

4 + 3 = 7

2.

_____

3.

_____

4.

_____

5.

_____

6.

_____

Practice 4

Add the numbers together by counting the vegetables. Write the answer on the pot.

1. **3 + 4 =**

  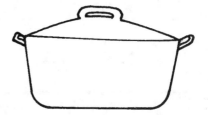

2. **2 + 3 =**

  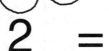 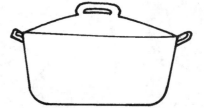

3. **4 + 2 =**

  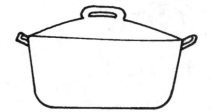

4. **3 + 5 =**

  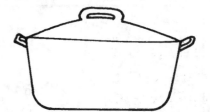

5. **2 + 2 =**

   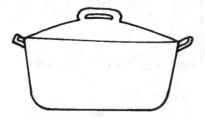

6. **4 + 5 =**

Practice 5

Write the sums.

1.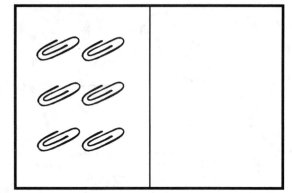

$$6 + 0 = \underline{\quad}$$

2.

$$4 + 3 = \underline{\quad}$$

3.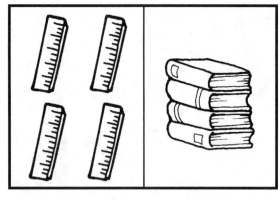

$$4 + 4 = \underline{\quad}$$

4.

$$2 + 2 = \underline{\quad}$$

5.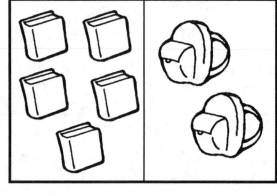

$$5 + 2 = \underline{\quad}$$

6.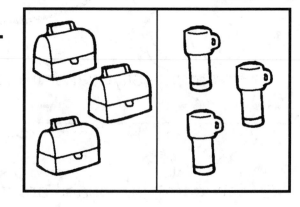

$$3 + 3 = \underline{\quad}$$

Practice 6

Color the jellybeans.

3 red	1 purple	5 orange	8 black
2 brown	4 pink	6 green	7 yellow

1. How many jellybeans are red or green? _____ + _____ = _____
    red green

2. How many jellybeans are orange or pink? _____ + _____ = _____
    orange pink

3. How many jellybeans are yellow or brown? _____ + _____ = _____
    yellow brown

4. How many jellybeans are purple or black? _____ + _____ = _____
    purple black

Practice 7

Guess what is in the box. Find the sums. Then write the letter in each box that matches each sum. Read the word you spelled out and draw it in the box.

8	9	10	11	12	13
s	e	a	o	m	u

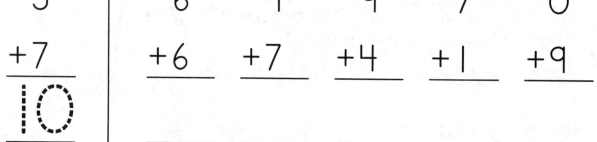

$$3 \quad\quad 6 \quad\quad 4 \quad\quad 9 \quad\quad 7 \quad\quad 0$$
$$+7 \quad +6 \quad +7 \quad +4 \quad +1 \quad +9$$

10

a

Practice 8

Cross out each answer in the net as you solve the problems.

1. $0 + 5 =$

2. $1 + 7 =$

3. $3 + 8 =$

4. $4 + 2 =$

5. $5 + 10 =$

6. $6 + 7 =$

7. $7 + 3 =$

8. $8 + 8 =$

9. $9 + 6 =$

5 8 13 6 11

15 10 16 15

#3315 Practice Makes Perfect: Addition and Subtraction

Practice 9

Cross out each answer in the balloons as you solve the problems.

1. 0 + 6 =

2. 1 + 10 =

3. 2 + 3 =

4. 3 + 5 =

5. 4 + 7 =

6. 5 + 9 =

7. 6 + 8 =

8. 7 + 4 =

9. 8 + 2 =

10. 9 + 5 =

11. 10 + 9 =

Practice 10

Guess what is in the box. Find the sums. Then write the letter in each box that matches each sum. Read the word you spelled out and draw it in the box.

10	11	12	13	14	15	16
y	e	a	o	m	n	k

5	8	6	8	10	2	5
+7	+6	+7	+7	+6	+9	+5

12
a

Practice 11

Find the sums.

1. 9 + 8 =	11. 5 + 7 =	21. 4 + 10 =
2. 7 + 9 =	12. 6 + 0 =	22. 7 + 1 =
3. 10 + 3 =	13. 7 + 6 =	23. 7 + 2 =
4. 6 + 6 =	14. 8 + 1 =	24. 5 + 5 =
5. 0 + 3 =	15. 5 + 4 =	25. 5 + 6 =
6. 8 + 5 =	16. 9 + 7 =	26. 8 + 6 =
7. 10 + 7 =	17. 10 + 6 =	27. 3 + 9 =
8. 0 + 10 =	18. 6 + 10 =	28. 9 + 2 =
9. 1 + 5 =	19. 9 + 0 =	29. 10 + 10 =
10. 2 + 0 =	20. 4 + 9 =	30. 4 + 0 =

Practice 12

Find the sums.

1. $1 + 0 =$	11. $10 + 1 =$	21. $1 + 1 =$
2. $4 + 4 =$	12. $1 + 2 =$	22. $9 + 9 =$
3. $0 + 9 =$	13. $6 + 1 =$	23. $3 + 6 =$
4. $10 + 0 =$	14. $7 + 5 =$	24. $8 + 0 =$
5. $8 + 7 =$	15. $1 + 6 =$	25. $2 + 9 =$
6. $0 + 4 =$	16. $1 + 9 =$	26. $6 + 2 =$
7. $9 + 10 =$	17. $2 + 10 =$	27. $9 + 1 =$
8. $7 + 8 =$	18. $3 + 0 =$	28. $0 + 8 =$
9. $8 + 10 =$	19. $3 + 1 =$	29. $9 + 4 =$
10. $5 + 0 =$	20. $0 + 0 =$	30. $7 + 10 =$

Practice 13

Use the pictures to solve these problems.

1.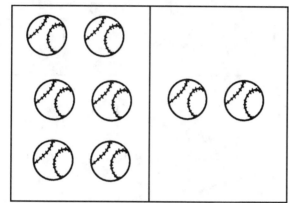

$$\underline{}6\underline{} - \underline{}2\underline{} = \underline{}$$

2.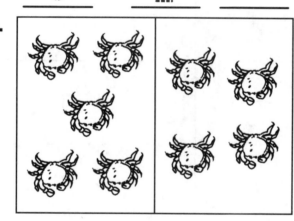

$$\underline{} - \underline{} = \underline{}$$

3.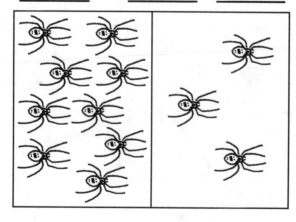

$$\underline{} - \underline{} = \underline{}$$

4.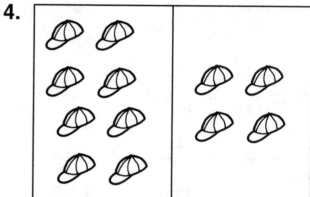

$$\underline{} - \underline{} = \underline{}$$

5.

$$\underline{} - \underline{} = \underline{}$$

6.

$$\underline{} - \underline{} = \underline{}$$

#3315 *Practice Makes Perfect: Addition and Subtraction*

Practice 14 ๑ ๏ ๑ ๏ ๑ ๏ ๑ ๏ ๑ ๏ ๑ ๏ ๑ ๏

Use the pictures to solve these problems.

1.

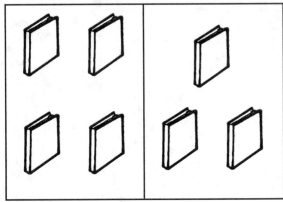

$$\text{4} \quad - \quad \text{3} \quad = \quad \underline{\quad}$$

2.

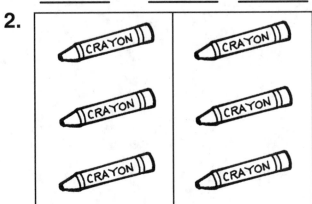

$$\underline{\quad} \quad - \quad \underline{\quad} \quad = \quad \underline{\quad}$$

3.

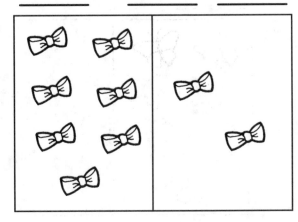

$$\underline{\quad} \quad - \quad \underline{\quad} \quad = \quad \underline{\quad}$$

4.

$$\underline{\quad} \quad - \quad \underline{\quad} \quad = \quad \underline{\quad}$$

5.

$$\underline{\quad} \quad - \quad \underline{\quad} \quad = \quad \underline{\quad}$$

6.

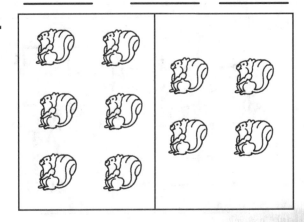

$$\underline{\quad} \quad - \quad \underline{\quad} \quad = \quad \underline{\quad}$$

Practice 15

Use the pictures to solve these problems.

1.

$$\underline{}\;-\;\underline{}\;=\;\underline{}$$

2.

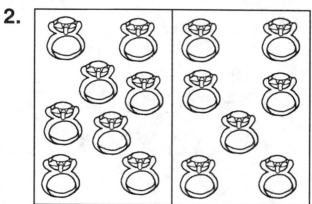

$$\underline{}\;-\;\underline{}\;=\;\underline{}$$

3.

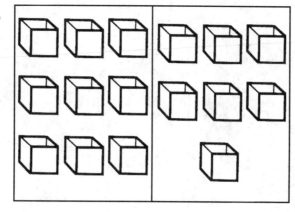

$$\underline{}\;-\;\underline{}\;=\;\underline{}$$

4.

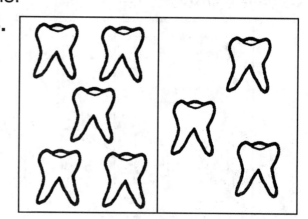

$$\underline{}\;-\;\underline{}\;=\;\underline{}$$

5.

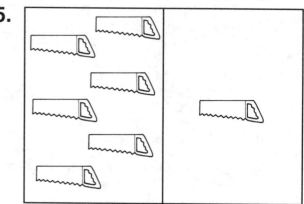

$$\underline{}\;-\;\underline{}\;=\;\underline{}$$

6.

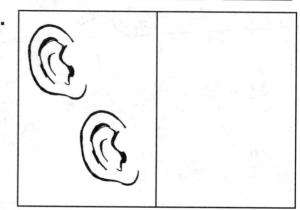

$$\underline{}\;-\;\underline{}\;=\;\underline{}$$

#3315 Practice Makes Perfect: Addition and Subtraction

Practice 16

Solve each problem below.

1. Cross out 4 sandwiches.

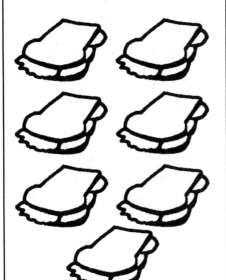

How many are left?_____

2. Cross out 2 peanuts.

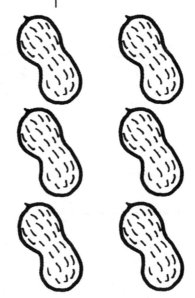

How many are left?_____

3. Cross out 1 pineapple.

How many are left?_____

4. Cross out 5 worms.

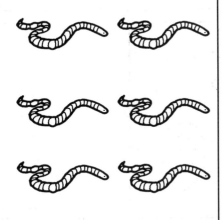

How many are left?_____

5. Cross out 3 shells.

How many are left?_____

6. Cross out 0 clocks.

How many are left?_____

Practice 17

Subtraction is counting backwards. Put the kangaroo back one jump on each number line. Write the number where she stops. The first one has been done for you.

1. 5 – 1 = ___4___

2. 4 – 1 = _____
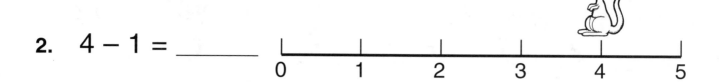

3. 3 – 1 = _____
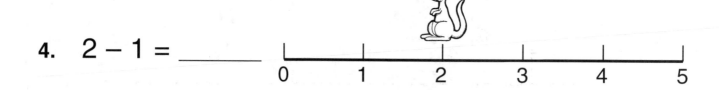

4. 2 – 1 = _____
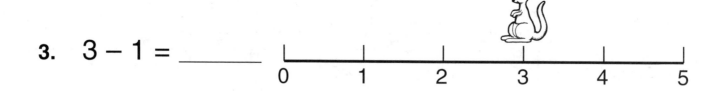

5. 1 – 1 = _____
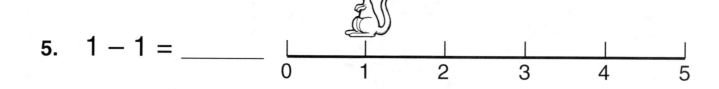

Practice 18

Follow the frog as he hops along the number line.

1. How many hops did the frog hop?   6 – 3 =

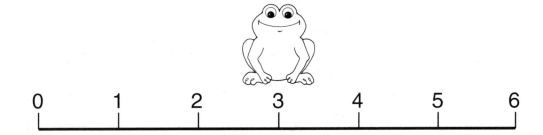

2. How many hops did the frog hop?   6 – 2 =

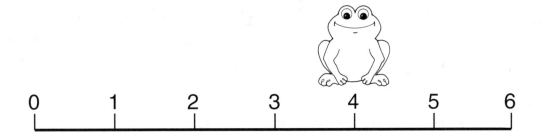

3. How many hops did the frog hop?   6 – 4 =

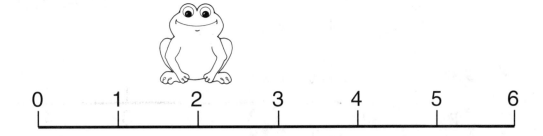

4. How many hops did the frog hop?   6 – 5 =

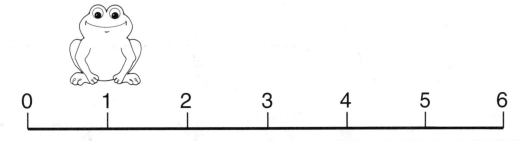

Practice 19

1. Count the balloons. Cross out 3 balloons. Write the numerals on the lines to show the subtraction problem and answer.

_____ – _____ = _____

2. Count the monkeys. Cross out 5 monkeys. Write the numerals on the lines to show the subtraction problem and answer.

_____ – _____ = _____

3. Count the mice. Cross out 2 mice. Write the numerals on the lines to show the subtraction problem and answer.

_____ – _____ = _____

4. Count the bees. Cross out 4 bees. Write the numerals on the lines to show the subtraction problem and answer.

_____ – _____ = _____

5. Count the hands. Cross out 0 hands. Write the numerals on the lines to show the subtraction problem and answer.

_____ – _____ = _____

Practice 20

1. Count the apples. Cross out 4 apples. Write the numerals on the lines to show the subtraction problem and answer.

_____ – _____ = _____

2. Count the baskets. Cross out 2 baskets. Write the numerals on the lines to show the subtraction problem and answer.

_____ – _____ = _____

3. Count the pens. Cross out 3 pens. Write the numerals on the lines to show the subtraction problem and answer.

_____ – _____ = _____

4. Count the chicks. Cross out 4 chicks. Write the numerals on the lines to show the subtraction problem and answer.

_____ – _____ = _____

5. Count the snakes. Cross out 1 snake. Write the numerals on the lines to show the subtraction problem and answer.

_____ – _____ = _____

Practice 21

Cross out each answer in the crayon box as you solve the problems.

1.
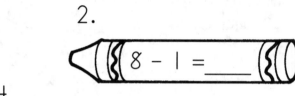
$1 - 0 = $ ____

2.
$8 - 1 = $ ____

3.
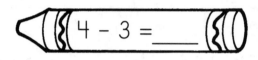
$5 - 2 = $ ____

4.
$4 - 3 = $ ____

5.
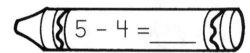
$8 - 4 = $ ____

6.
$5 - 4 = $ ____

7.
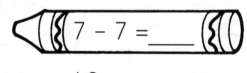
$9 - 6 = $ ____

8.
$7 - 7 = $ ____

9.
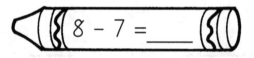
$8 - 7 = $ ____

11.
$9 - 3 = $ ____

10.
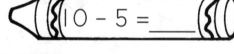
$10 - 5 = $ ____

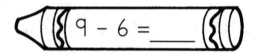

My Crayons

| 7 | 3 | 1 | 1 | 4 |
| 3 | 1 | 0 | 6 | 5 | 1 |

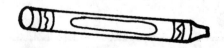 

Practice 22

Guess what is in the box. Solve each problem. Then write the letter in each box that matches each answer. Read the word you spelled out and draw it in the box.

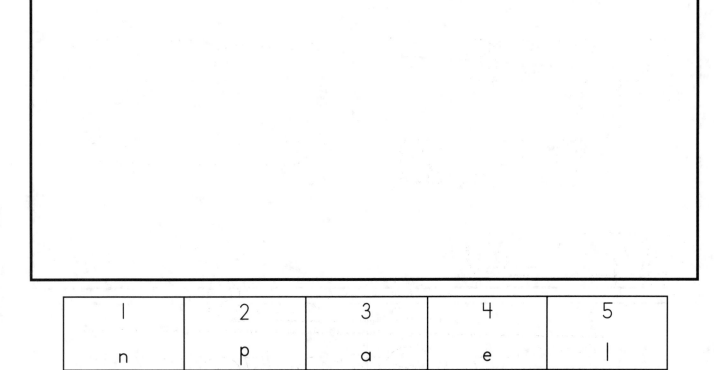

1	2	3	4	5
n	p	a	e	l

6	5		4	6	5	10	9
−3	−4		−1	−4	−3	−5	−5
3							
a	☐		☐	☐	☐	☐	☐

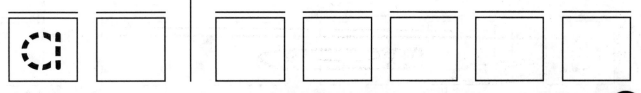

Practice 23

Guess what is in the box. Solve each problem. Then write the letter in each box that matches each answer. Read the word you spelled out and draw it in the box.

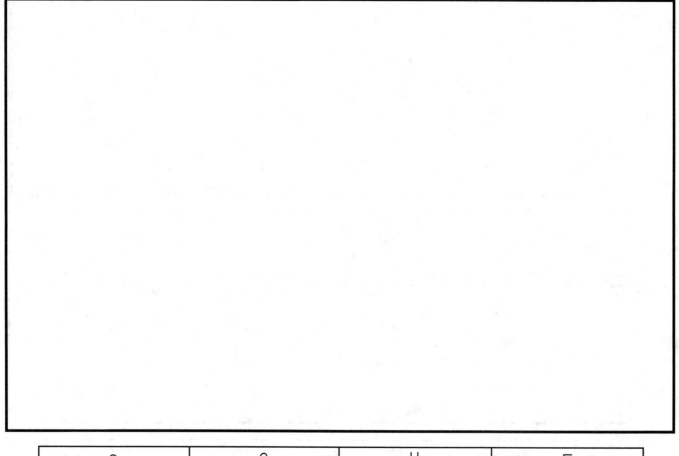

2	3	4	5
p	u	a	y

$$\begin{array}{r} 7 \\ -3 \\ \hline 4 \end{array}$$

$$\begin{array}{r} 6 \\ -4 \\ \hline \end{array}$$

$$\begin{array}{r} 7 \\ -4 \\ \hline \end{array}$$

$$\begin{array}{r} 9 \\ -7 \\ \hline \end{array}$$

$$\begin{array}{r} 3 \\ -1 \\ \hline \end{array}$$

$$\begin{array}{r} 8 \\ -3 \\ \hline \end{array}$$

a

Practice 24

Guess what is in the box. Solve each problem. Then write the letter in each box that matches each answer. Read the word you spelled out and draw it in the box.

0	1	2	3	4	5
l	f	a	o	b	t

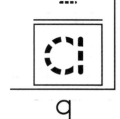

7	8	9	3	6
-5	-7	-6	-0	-1

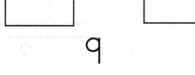 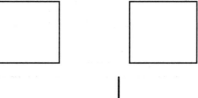 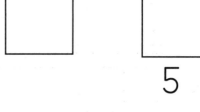

9	9	1	5
-5	-7	-1	-5

Practice 25

Find the answers.

10. $2 - 0 =$ ___

20. $9 - 4 =$ ___

30. $4 - 0 =$ ___

1. $9 - 8 =$ ___

11. $7 - 5 =$ ___

21. $10 - 4 =$ ___

31. $1 - 0 =$ ___

2. $9 - 7 =$ ___

12. $6 - 0 =$ ___

22. $7 - 1 =$ ___

32. $4 - 4 =$ ___

3. $10 - 3 =$ ___

13. $7 - 6 =$ ___

23. $7 - 2 =$ ___

33. $9 - 6 =$ ___

4. $6 - 6 =$ ___

14. $8 - 1 =$ ___

24. $5 - 5 =$ ___

34. $10 - 0 =$ ___

5. $3 - 0 =$ ___

15. $5 - 4 =$ ___

25. $6 - 5 =$ ___

35. $8 - 5 =$ ___

6. $8 - 2 =$ ___

16. $9 - 3 =$ ___

26. $8 - 6 =$ ___

36. $4 - 1 =$ ___

7. $10 - 7 =$ ___

17. $10 - 6 =$ ___

27. $9 - 1 =$ ___

37. $2 - 2 =$ ___

8. $10 - 9 =$ ___

18. $6 - 2 =$ ___

28. $9 - 2 =$ ___

38. $8 - 7 =$ ___

9. $5 - 1 =$ ___

19. $9 - 0 =$ ___

29. $10 - 10 =$ ___

39. $10 - 8 =$ ___

Practice 26

Subtract. Write how many are left.

1.

$$\begin{array}{r} 5 \\ -\ 1 \\ \hline \end{array}$$

4

2.

$$\begin{array}{r} 6 \\ -\ 4 \\ \hline \end{array}$$

3.

$$\begin{array}{r} 3 \\ -\ 1 \\ \hline \end{array}$$

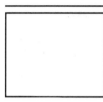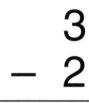

4.

$$\begin{array}{r} 3 \\ -\ 2 \\ \hline \end{array}$$

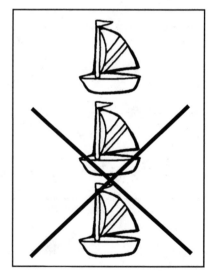

5.

$$\begin{array}{r} 4 \\ -\ 1 \\ \hline \end{array}$$

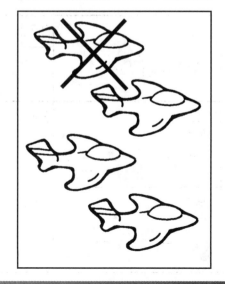

6.

$$\begin{array}{r} 2 \\ -\ 0 \\ \hline \end{array}$$

Practice 27

Finish the subtraction sentences. Write the answer. Draw the number of dots.

1. 6 dots take away 4 dots = _____ dots

2. 6 dots take away 2 dots = _____ dots

3. 6 dots take away 1 dot = _____ dots

4. 6 dots take away 3 dots = _____ dots

5. 6 dots take away 5 dots = _____ dot

6. 6 dots take away 6 dots = _____ dots

Practice 28

1. $\begin{array}{r} 3 \\ -\ 1 \\ \hline \end{array}$	2. $\begin{array}{r} 4 \\ -\ 2 \\ \hline \end{array}$	3. $\begin{array}{r} 2 \\ -\ 2 \\ \hline \end{array}$	4. $\begin{array}{r} 5 \\ -\ 3 \\ \hline \end{array}$	5. $\begin{array}{r} 3 \\ -\ 3 \\ \hline \end{array}$
6. $\begin{array}{r} 5 \\ -\ 1 \\ \hline \end{array}$	7. $\begin{array}{r} 3 \\ -\ 2 \\ \hline \end{array}$	8. $\begin{array}{r} 4 \\ -\ 3 \\ \hline \end{array}$	9. $\begin{array}{r} 5 \\ -\ 4 \\ \hline \end{array}$	10. $\begin{array}{r} 2 \\ -\ 1 \\ \hline \end{array}$
11. $\begin{array}{r} 4 \\ -\ 4 \\ \hline \end{array}$	12. $\begin{array}{r} 5 \\ -\ 2 \\ \hline \end{array}$	13. $\begin{array}{r} 1 \\ -\ 1 \\ \hline \end{array}$	14. $\begin{array}{r} 5 \\ -\ 5 \\ \hline \end{array}$	15. $\begin{array}{r} 4 \\ -\ 1 \\ \hline \end{array}$
16. $\begin{array}{r} 2 \\ -\ 1 \\ \hline \end{array}$	17. $\begin{array}{r} 1 \\ -\ 1 \\ \hline \end{array}$	18. $\begin{array}{r} 3 \\ -\ 2 \\ \hline \end{array}$	19. $\begin{array}{r} 5 \\ -\ 4 \\ \hline \end{array}$	20. $\begin{array}{r} 4 \\ -\ 3 \\ \hline \end{array}$

Practice 29

Read each word problem. Write the number sentence it shows. Find the difference.

1.	2.
A dog was walking through his yard when he saw 5 cats on the fence. He barked, and 2 cats ran away. How many cats were left?	 A woman grew a vegetable garden, and in it there were 10 ears of corn. She picked 3 ears on one day, 2 ears on the next, and 4 ears on the third day. How many ears of corn were left in her garden?
$5 - 2 = 3$	

3.	4.
 A monkey had 10 bananas. He ate 3 the first day and 2 the second. How many bananas were left?	 Andrew had 6 assignments for homework. He completed 2 before dinner and 2 after dinner. How many assignments were left?

Practice 30

To discover the secret shape, follow the directions. Find the sums. Then connect the answers with a line. Connect the first answer with the second answer and the second answer with the third answer. Continue until you finish the shape.

1.

$$17 + 22$$

2.

$$54 + 21$$

3.

$$76 + 22$$

4.

$$18 + 31$$

5.

$$70 + 20$$

6.

$$11 + 11$$

7.

$$51 + 27$$

8.

$$19 + 10$$

9.

$$62 + 22$$

10.

$$35 + 21$$

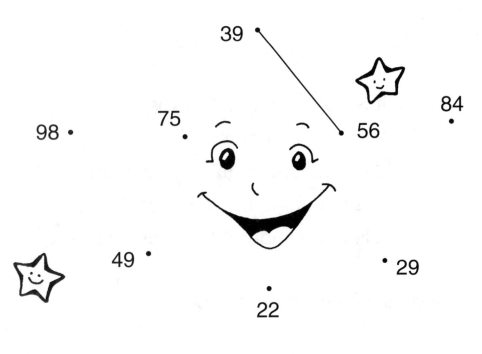

Practice 31

To discover the secret number, find the sums and follow the directions.

1.

```
   21
 + 18
 ─────
```

2.

```
   31
 + 16
 ─────
```

3.

```
   31
 + 21
 ─────
```

4.

```
   41
 + 31
 ─────
```

5.

```
   12
 + 12
 ─────
```

6.

```
   10
 + 17
 ─────
```

☆ It is not number 24. Cross it out.

☆ It is not number 39. Cross it out.

☆ It is not number 52. Cross it out.

☆ It is not number 72. Cross it out.

☆ It is not number 47. Cross it out.

What is the secret number? _____

Practice 32

Solve the problems on each shape. Can you name each one?

Color the shapes that you can identify.

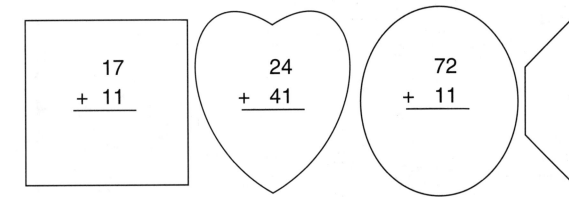

1.

17
+ 11

2.

24
+ 41

3.

72
+ 11

4.

10
+ 10

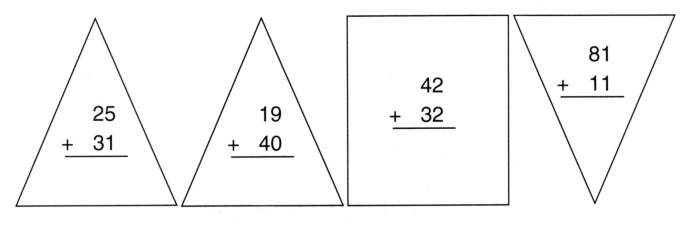

5.

25
+ 31

6.

19
+ 40

7.

42
+ 32

8.

81
+ 11

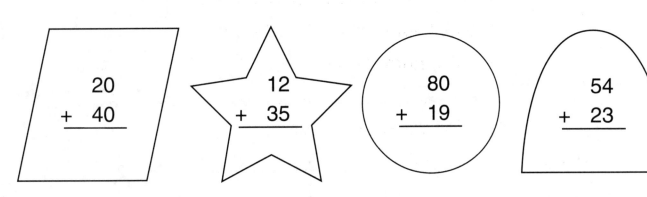

9.

20
+ 40

10.

12
+ 35

11.

80
+ 19

12.

54
+ 23

Practice 33

Solve the problems below.

1.
$$\begin{array}{r} 29 \\ -19 \\ \hline \end{array}$$

2.
$$\begin{array}{r} 68 \\ -23 \\ \hline \end{array}$$

3.
$$\begin{array}{r} 77 \\ -33 \\ \hline \end{array}$$

4.
$$\begin{array}{r} 67 \\ -14 \\ \hline \end{array}$$

5.
$$\begin{array}{r} 79 \\ -20 \\ \hline \end{array}$$

6.
$$\begin{array}{r} 68 \\ -26 \\ \hline \end{array}$$

7.
$$\begin{array}{r} 39 \\ -24 \\ \hline \end{array}$$

8.
$$\begin{array}{r} 39 \\ -20 \\ \hline \end{array}$$

9.
$$\begin{array}{r} 59 \\ -23 \\ \hline \end{array}$$

10.
$$\begin{array}{r} 58 \\ -40 \\ \hline \end{array}$$

11.
$$\begin{array}{r} 29 \\ -12 \\ \hline \end{array}$$

12.
$$\begin{array}{r} 89 \\ -21 \\ \hline \end{array}$$

13.
$$\begin{array}{r} 49 \\ -23 \\ \hline \end{array}$$

14.
$$\begin{array}{r} 78 \\ -32 \\ \hline \end{array}$$

15.
$$\begin{array}{r} 68 \\ -41 \\ \hline \end{array}$$

Practice 34

Below is a famous saying about birds. Solve the problems in each box below. Then find the answer under the word blank and place the letter from the box on the matching blank.

| ‾‾ | ‾‾ | ‾‾ | ‾‾ | ‾‾ | | ‾‾ | ‾‾ | ‾‾ | ‾‾ |
| 32 | 13 | 43 | 38 | 13 | | 38 | 35 | 87 | 29 |

| ‾‾ | ‾‾ | ‾‾ | ‾‾ | ‾‾ | | ‾‾ | ‾‾ | ‾‾ |
| 21 | 43 | 10 | 48 | 33 | , | 33 | 13 | 29 |

| ‾‾ | ‾‾ | ‾‾ | ‾‾ | ‾‾ | ‾‾ | ‾‾ | | ‾‾ | ‾‾ |
| 38 | 13 | 43 | 38 | 65 | 29 | 64 | | 1 | 10 |

| ‾‾ | ‾‾ | ‾‾ | | ‾‾ | ‾‾ | ‾‾ |
| 33 | 13 | 29 | | 29 | 59 | 59 | ?

O 16 − 15	**H** 26 − 13	**T** 30 + 3	**W** 42 − 10	**E** 22 + 7
I 67 − 24	**F** 38 − 17	**G** 48 + 11	**A** 22 + 13	**C** 15 + 23
M 63 + 24	**R** 42 − 32	**N** 68 − 4	**S** 23 + 25	**K** 89 − 24

I think the _____ came first because _____

_____

_____

Practice 35

Fill in the missing number on each cone to complete the problem.

1.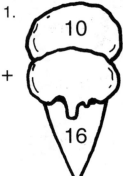
$+$ 10 / 16

2.
$-$ 17 / 8

3.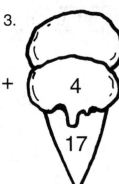
$+$ 4 / 17

4.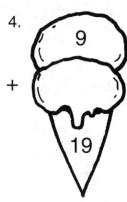
$+$ 9 / 19

5.
$-$ 8 / 1

6.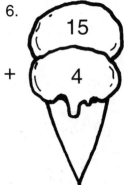
$+$ 15 / 4

7.
$-$ 11 / 7

8.
$-$ 21 / 16

9.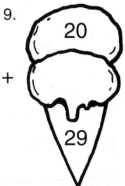
$+$ 20 / 29

10.
$-$ 14 / 7

11.
$-$ 13 / 10

12.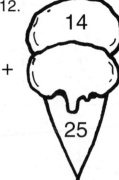
$+$ 14 / 25

13.
$-$ 14 / 6

14.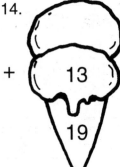
$+$ 13 / 19

15.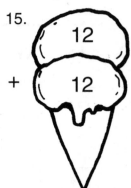
$+$ 12 / 12

16.
$-$ 18 / 6

Practice 36

1. There are 5 🐔's on the farm.

Each 🐔 laid 3 ◯'s.

How many ◯'s are there

altogether? _____

2. Seven beautiful 🌸's are

growing in the field. You pick 4

of the 🌸's. How many are

left? _____

3. One apple tree has 7 🍎's.

The other tree has 6 🍎's.

You eat 2. How many

apples are left? _____

4. On the farm, there are 3 🐕's,

5 🐤, 2 🐱, and

2 🐄's. What is the total

number of animals on the

farm? _____

5. The farmer wants to have

10 🌳's in the orchard.

He now has 3. How many

more 🌳's will he need to

plant? _____

6. Three 🐄's are standing

near the barn. Each cow gives

3 🪣's of milk. How many

🪣's of milk? _____

Test Practice 1 ＠＠＠＠＠＠＠＠＠＠＠＠

Solve each problem and fill in the correct circle that matches each answer.

1.

4 + 5 = _____

Ⓐ 9

Ⓑ 4

Ⓒ 6

Ⓓ 10

4.

5 + 8 = _____

Ⓐ 12

Ⓑ 13

Ⓒ 11

Ⓓ 10

2.

6 + 7 = _____

Ⓐ 11

Ⓑ 5

Ⓒ 12

Ⓓ 13

5.

10 + 12 = _____

Ⓐ 12

Ⓑ 24

Ⓒ 22

Ⓓ 23

3.

2 + 9 = _____

Ⓐ 14

Ⓑ 10

Ⓒ 11

Ⓓ 8

6.

3 + 11 = _____

Ⓐ 14

Ⓑ 16

Ⓒ 24

Ⓓ 21

Test Practice 2

Solve each problem and fill in the correct circle that matches each answer.

1.

$11 + 17 = \square$

(A) 19

(B) 21

(C) 20

(D) 28

4.

$9 + 13 = \square$

(A) 13

(B) 22

(C) 10

(D) 11

2.

$\square + 4 = 15$

(A) 18

(B) 7

(C) 9

(D) 11

5.

$\square + 9 = 18$

(A) 7

(B) 6

(C) 9

(D) 8

3.

$12 + \square = 22$

(A) 6

(B) 10

(C) 9

(D) 7

6.

$15 + \square = 27$

(A) 7

(B) 10

(C) 5

(D) 12

Test Practice 3 ⟩ ❧ ⟩ ⟩ ❧ ⟩ ⟩ ❧ ⟩ ⟩ ❧ ⟩ ⟩ ❧

Solve each problem and fill in the correct circle that matches each answer.

1.

10 − 8 = _____

(A) 0

(B) 1

(C) 2

(D) 3

4.

17 − 12 = _____

(A) 9

(B) 5

(C) 6

(D) 8

2.

7 − 4 = _____

(A) 3

(B) 11

(C) 10

(D) 4

5.

15 − 2 = _____

(A) 10

(B) 12

(C) 13

(D) 14

3.

9 − 1 = _____

(A) 7

(B) 6

(C) 5

(D) 8

6.

11 − 8 = _____

(A) 3

(B) 2

(C) 1

(D) 4

Test Practice 4 ⟐ ⟐ ⟐ ⟐ ⟐ ⟐ ⟐ ⟐ ⟐ ⟐ ⟐

Solve each problem and fill in the correct circle that matches each answer.

1.

$35 - 10 = \square$

Ⓐ 15

Ⓑ 20

Ⓒ 25

Ⓓ 30

4.

$26 - 8 = \square$

Ⓐ 12

Ⓑ 34

Ⓒ 18

Ⓓ 20

2.

$22 - \square = 15$

Ⓐ 11

Ⓑ 8

Ⓒ 9

Ⓓ 7

5.

$27 - \square = 13$

Ⓐ 24

Ⓑ 14

Ⓒ 40

Ⓓ 30

3.

$\square - 14 = 30$

Ⓐ 24

Ⓑ 44

Ⓒ 34

Ⓓ 54

6.

$\square - 16 = 23$

Ⓐ 23

Ⓑ 39

Ⓒ 29

Ⓓ 38

Test Practice 5

Solve each problem and fill in the correct circle that matches each answer.

1.

$$\begin{array}{r} 5 \\ 8 \\ + 7 \\ \hline \end{array}$$

Ⓐ 22
Ⓑ 20
Ⓒ 21
Ⓓ 23

4.

$$\begin{array}{r} 3 \\ 9 \\ + 1 \\ \hline \end{array}$$

Ⓐ 12
Ⓑ 15
Ⓒ 13
Ⓓ 14

2.

$$\begin{array}{r} 2 \\ 8 \\ + 4 \\ \hline \end{array}$$

Ⓐ 14
Ⓑ 11
Ⓒ 15
Ⓓ 10

5.

$$\begin{array}{r} 2 \\ 5 \\ + 1 \\ \hline \end{array}$$

Ⓐ 6
Ⓑ 7
Ⓒ 8
Ⓓ 9

3.

$$\begin{array}{r} 4 \\ 5 \\ + 6 \\ \hline \end{array}$$

Ⓐ 16
Ⓑ 26
Ⓒ 11
Ⓓ 15

6.

$$\begin{array}{r} 10 \\ 3 \\ + 4 \\ \hline \end{array}$$

Ⓐ 15
Ⓑ 18
Ⓒ 16
Ⓓ 17

 #3315 Practice Makes Perfect: Addition and Subtraction

Test Practice 6 ⟡ ⟡ ⟡ ⟡ ⟡ ⟡ ⟡ ⟡ ⟡ ⟡ ⟡ ⟡

Solve each problem and fill in the correct circle that matches each answer.

1.

$$
\begin{array}{r}
10 \\
2 \\
-1 \\
\hline
\end{array}
$$

Ⓐ 9

Ⓑ 11

Ⓒ 12

Ⓓ 7

4.

$$
\begin{array}{r}
24 \\
10 \\
-8 \\
\hline
\end{array}
$$

Ⓐ 7

Ⓑ 9

Ⓒ 6

Ⓓ 8

2.

$$
\begin{array}{r}
15 \\
5 \\
-7 \\
\hline
\end{array}
$$

Ⓐ 4

Ⓑ 8

Ⓒ 3

Ⓓ 1

5.

$$
\begin{array}{r}
20 \\
3 \\
-9 \\
\hline
\end{array}
$$

Ⓐ 11

Ⓑ 9

Ⓒ 12

Ⓓ 8

3.

$$
\begin{array}{r}
18 \\
4 \\
-6 \\
\hline
\end{array}
$$

Ⓐ 8

Ⓑ 10

Ⓒ 12

Ⓓ 14

6.

$$
\begin{array}{r}
30 \\
20 \\
-5 \\
\hline
\end{array}
$$

Ⓐ 5

Ⓑ 10

Ⓒ 20

Ⓓ 15

Answer Sheet

Test Practice 1
1. Ⓐ Ⓑ Ⓒ Ⓓ
2. Ⓐ Ⓑ Ⓒ Ⓓ
3. Ⓐ Ⓑ Ⓒ Ⓓ
4. Ⓐ Ⓑ Ⓒ Ⓓ
5. Ⓐ Ⓑ Ⓒ Ⓓ
6. Ⓐ Ⓑ Ⓒ Ⓓ

Test Practice 2
1. Ⓐ Ⓑ Ⓒ Ⓓ
2. Ⓐ Ⓑ Ⓒ Ⓓ
3. Ⓐ Ⓑ Ⓒ Ⓓ
4. Ⓐ Ⓑ Ⓒ Ⓓ
5. Ⓐ Ⓑ Ⓒ Ⓓ
6. Ⓐ Ⓑ Ⓒ Ⓓ

Test Practice 3
1. Ⓐ Ⓑ Ⓒ Ⓓ
2. Ⓐ Ⓑ Ⓒ Ⓓ
3. Ⓐ Ⓑ Ⓒ Ⓓ
4. Ⓐ Ⓑ Ⓒ Ⓓ
5. Ⓐ Ⓑ Ⓒ Ⓓ
6. Ⓐ Ⓑ Ⓒ Ⓓ

Test Practice 4
1. Ⓐ Ⓑ Ⓒ Ⓓ
2. Ⓐ Ⓑ Ⓒ Ⓓ
3. Ⓐ Ⓑ Ⓒ Ⓓ
4. Ⓐ Ⓑ Ⓒ Ⓓ
5. Ⓐ Ⓑ Ⓒ Ⓓ
6. Ⓐ Ⓑ Ⓒ Ⓓ

Test Practice 5
1. Ⓐ Ⓑ Ⓒ Ⓓ
2. Ⓐ Ⓑ Ⓒ Ⓓ
3. Ⓐ Ⓑ Ⓒ Ⓓ
4. Ⓐ Ⓑ Ⓒ Ⓓ
5. Ⓐ Ⓑ Ⓒ Ⓓ
6. Ⓐ Ⓑ Ⓒ Ⓓ

Test Practice 6
1. Ⓐ Ⓑ Ⓒ Ⓓ
2. Ⓐ Ⓑ Ⓒ Ⓓ
3. Ⓐ Ⓑ Ⓒ Ⓓ
4. Ⓐ Ⓑ Ⓒ Ⓓ
5. Ⓐ Ⓑ Ⓒ Ⓓ
6. Ⓐ Ⓑ Ⓒ Ⓓ

Answer Key

Page 4
1. 4
2. 2 + 7 = 9
3. 5 + 3 = 8
4. 4 + 3 = 7
5. 4 + 5 = 9
6. 6 + 3 = 9

Page 5
1. 4
2. 3 + 3 = 6
3. 4 + 6 = 10
4. 5 + 1 = 6
5. 0 + 7 = 7
6. 5 + 8 = 13

Page 6
1. 7
2. 3 + 5 = 8
3. 2 + 4 = 6
4. 5 + 5 = 10
5. 6 + 3 = 9
6. 4 + 1 = 5

Page 7
1. 7
2. 5
3. 6
4. 8
5. 4
6. 9

Page 8
1. 6
2. 7
3. 8
4. 4
5. 7
6. 6

Page 9
1. 3 + 6 = 9
2. 5 + 4 = 9
3. 7 + 2 = 9
4. 1 + 8 = 9

Page 10
10; a
12; m
11; o
13; u
8; s
9; e
a mouse

Page 11
1. 5
2. 8
3. 11
4. 6
5. 15
6. 13
7. 10
8. 16
9. 15

Page 12
1. 6
2. 11
3. 5
4. 8
5. 11
6. 14
7. 14
8. 11
9. 10
10. 14
11. 19

Page 13
12; a
14; m
13; o
15; n
16; k
11; e
10; y
a monkey

Page 14
1. 17
2. 16
3. 13
4. 12
5. 3
6. 13
7. 17
8. 10
9. 6
10. 2
11. 12
12. 6
13. 13
14. 9
15. 9
16. 16
17. 16
18. 16
19. 9
20. 13
21. 14
22. 8
23. 9
24. 10
25. 11
26. 14
27. 12
28. 11
29. 20
30. 4

Page 15
1. 1
2. 8
3. 9
4. 10
5. 15
6. 4
7. 19
8. 15
9. 18
10. 5
11. 11
12. 3
13. 7
14. 12
15. 7
16. 10
17. 12
18. 3
19. 4
20. 0
21. 2
22. 18
23. 9
24. 8
25. 11
26. 8
27. 10
28. 8
29. 13
30. 17

Page 16
1. 4
2. 5 − 4 = 1
3. 9 − 3 = 6
4. 8 − 4 = 4
5. 2 − 1 = 1
6. 7 − 5 = 2

Page 17
1. 1
2. 3 − 3 = 0
3. 7 − 2 = 5
4. 9 − 5 = 4
5. 10 − 3 = 7
6. 6 − 4 = 2

Page 18
1. 0
2. 8 − 7 = 1
3. 9 − 7 = 2
4. 5 − 3 = 2
5. 6 − 1 = 5
6. 2 − 0 = 2

Page 19
1. 3
2. 4
3. 3
4. 1
5. 4
6. 5

Page 20
1. 4
2. 3
3. 2
4. 1
5. 1

Page 21
1. 3
2. 4
3. 2
4. 1

Page 22
1. 6 − 3 = 3
2. 7 − 5 = 2
3. 9 − 2 = 7
4. 10 − 4 = 6
5. 5 − 0 = 5

Page 23
1. 9 − 4 = 5
2. 8 − 2 = 6
3. 7 − 3 = 4
4. 8 − 4 = 4
5. 8 − 1 = 7

Page 24
1. 1
2. 7
3. 3
4. 1
5. 4
6. 1
7. 3
8. 0
9. 1
10. 5
11. 6

Page 25
3; a
1; n
3; a
2; p
2; p
5; l
4; e
an apple

Page 26
4; a
2; p
3; u
2; p
2; p
5; y
a puppy

Answer Key ꞵ ꙮ ꞵ ꙮ ꞵ ꙮ ꞵ ꙮ ꞵ ꙮ ꞵ ꙮ ꞵ

Page 27
2; a
1; f
3; o
3; o
5; t
4; b
2; a
0; l
0; l
a football

Page 28
1. 1	21. 6
2. 2	22. 6
3. 7	23. 5
4. 0	24. 0
5. 3	25. 1
6. 6	26. 2
7. 3	27. 8
8. 1	28. 7
9. 4	29. 0
10. 2	30. 4
11. 2	31. 1
12. 6	32. 0
13. 1	33. 3
14. 7	34. 10
15. 1	35. 3
16. 6	36. 3
17. 4	37. 0
18. 4	38. 1
19. 9	39. 2
20. 5	

Page 29
1. 4
2. 2
3. 2
4. 1
5. 3
6. 2

Page 30
1. 2
2. 4
3. 5
4. 3
5. 1
6. 0

Page 31
1. 2
2. 2
3. 0

4. 2
5. 0
6. 4
7. 1
8. 1
9. 1
10. 1
11. 0
12. 3
13. 0
14. 0
15. 3
16. 1
17. 0
18. 1
19. 1
20. 1

Page 32
1. 5 − 2 = 3
2. 10 − 3 − 2 − 4 = 1
3. 10 − 3 − 2 = 5
4. 6 − 2 − 2 = 2

Page 33
1. 39
2. 75
3. 98
4. 49
5. 90
6. 22
7. 78
8. 29
9. 84
10. 56

Page 34
1. 39
2. 47
3. 52
4. 72
5. 24
6. 27

secret number: 27

Page 35
1. 28	7. 74
2. 65	8. 92
3. 83	9. 60
4. 20	10. 47
5. 56	11. 99
6. 59	12. 77

Page 36
1. 10
2. 45
3. 44
4. 53
5. 59
6. 42
7. 15
8. 19
9. 36
10. 18
11. 17
12. 68
13. 26
14. 46
15. 27

Page 37
O = 1
H = 13
T = 33
W = 32
E = 29
I = 43
F = 21
G = 59
A = 35
C = 38
M = 87
R = 10
N = 64
S = 48
K = 65
famous saying: Which came first, the chicken or the egg?

Page 38
1. 6
2. 9
3. 13
4. 10
5. 7
6. 19
7. 18
8. 5
9. 9
10. 7
11. 23
12. 11
13. 8
14. 6
15. 24
16. 12

Page 39
1. 15
2. 3
3. 11
4. 12
5. 7
6. 9

Page 40
1. A
2. D
3. C
4. B
5. C
6. A

Page 41
1. D
2. D
3. B
4. B
5. C
6. D

Page 42
1. C
2. A
3. D
4. B
5. C
6. A

Page 43
1. C
2. D
3. B
4. C
5. B
6. B

Page 44
1. B
2. A
3. D
4. C
5. C
6. D

Page 45
1. D
2. C
3. A
4. C
5. D
6. A